Math Practice Sheet

600 Multiplications

4 Digit by 3 Digit

$$\begin{array}{r} 5279 \\ \times \quad 862 \\ \hline \end{array}$$

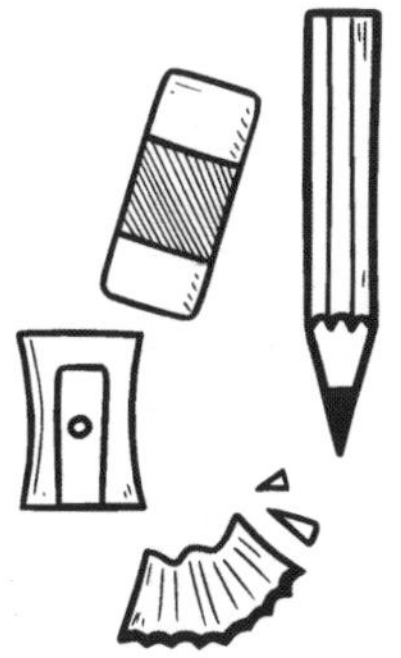

Name: ______________________ Date: __________

Multiplying in Columns

Directions: Solve the Multiplications

7054 × 258	5840 × 324	9936 × 846	2362 × 375
5095 × 533	4695 × 635	3627 × 275	5279 × 862
4326 × 492	4548 × 508	8869 × 442	6263 × 520

Name: ________________________ Date: ____________

Multiplying in Columns

Directions: Solve the Multiplications

4414 × 221	5369 × 580	8794 × 587	2309 × 203
1390 × 120	8575 × 324	7502 × 202	4102 × 901
6309 × 452	2540 × 401	7708 × 542	9852 × 382

Name: ______________________________ Date: __________

Multiplying in Columns

Directions: Solve the Multiplications

2030 × 401	8021 × 502	5320 × 213	4521 × 784
4579 × 124	1247 × 231	4587 × 245	3568 × 702
4582 × 785	7845 × 582	6582 × 365	5872 × 652

Name: ______________________ Date: ____________ 04

Multiplying in Columns

Directions: Solve the Multiplications

4211 × 221	2231 × 121	8790 × 582	9987 × 120
4521 × 782	5874 × 201	7802 × 320	5601 × 218
4785 × 120	7854 × 952	7845 × 854	7894 × 321

Name: ______________________ Date: ____________

Multiplying in Columns

Directions: Solve the Multiplications

2140 × 324	4253 × 325	9780 × 731	3870 × 328
9802 × 127	3278 × 281	2397 × 378	3879 × 387
7892 × 287	7032 × 302	3087 × 103	7821 × 108

Name: ____________________ Date: __________ 06

Multiplying in Columns

Directions: Solve the Multiplications

$\begin{array}{r} 9870 \\ \times \quad 907 \\ \hline \end{array}$	$\begin{array}{r} 9748 \\ \times \quad 321 \\ \hline \end{array}$	$\begin{array}{r} 8974 \\ \times \quad 839 \\ \hline \end{array}$	$\begin{array}{r} 7854 \\ \times \quad 385 \\ \hline \end{array}$
$\begin{array}{r} 8965 \\ \times \quad 385 \\ \hline \end{array}$	$\begin{array}{r} 7823 \\ \times \quad 287 \\ \hline \end{array}$	$\begin{array}{r} 8952 \\ \times \quad 127 \\ \hline \end{array}$	$\begin{array}{r} 1237 \\ \times \quad 820 \\ \hline \end{array}$
$\begin{array}{r} 7838 \\ \times \quad 472 \\ \hline \end{array}$	$\begin{array}{r} 7841 \\ \times \quad 287 \\ \hline \end{array}$	$\begin{array}{r} 6451 \\ \times \quad 174 \\ \hline \end{array}$	$\begin{array}{r} 4650 \\ \times \quad 465 \\ \hline \end{array}$

Name: ____________________ Date: __________

Multiplying in Columns

Directions: Solve the Multiplications

4244 × 223	4548 × 321	4785 × 235	4578 × 369
4369 × 321	7896 × 314	9630 × 369	3795 × 103
6324 × 326	9782 × 901	4870 × 701	9870 × 710

Name: ____________________ Date: __________

Multiplying in Columns

Directions: Solve the Multiplications

5820 × 320	3690 × 105	2017 × 301	9780 × 504
7894 × 321	8974 × 963	7845 × 401	4014 × 789
9870 × 254	7485 × 782	7890 × 985	9685 × 785

Name: ____________________ Date: __________

Multiplying in Columns

Directions: Solve the Multiplications

7412 × 328	9632 × 963	7410 × 698	4789 × 632
7845 × 852	8745 × 658	9865 × 521	7854 × 724
6985 × 758	3570 × 782	7458 × 740	7458 × 750

Name: ____________________ Date: __________

Multiplying in Columns

Directions: Solve the Multiplications

7080 × 250	7854 × 235	7835 × 230	3285 × 421
4780 × 385	3685 × 257	7845 × 325	3980 × 247
3985 × 860	6080 × 782	7845 × 547	9865 × 740

Name: ______________________ Date: __________

Multiplying in Columns

Directions: Solve the Multiplications

7850 × 253	7854 × 325	7845 × 369	7845 × 982
9632 × 358	9850 × 441	2583 × 258	3639 × 741
2587 × 369	7894 × 987	4563 × 654	1230 × 321

Name: ______________________ Date: ____________

Multiplying in Columns

Directions: Solve the Multiplications

3214 × 456	3247 × 236	7896 × 208	3698 × 963
2587 × 852	1478 × 741	9632 × 789	7834 × 247
7821 × 324	7896 × 378	1479 × 327	9290 × 923

Name: ______________________________ Date: ______________

Multiplying in Columns

Directions: Solve the Multiplications

3240 × 258	2170 × 210	7845 × 120	1750 × 124
2010 × 741	7879 × 201	7840 × 985	1202 × 283
3695 × 147	7852 × 412	7845 × 325	7845 × 124

Name: ______________________________ Date: ____________

Multiplying in Columns

Directions: Solve the Multiplications

7852 × 481	6985 × 241	9874 × 251	8745 × 321
7845 × 214	3258 × 124	3258 × 784	9852 × 324
3692 × 475	7845 × 214	7841 × 782	7896 × 276

Name: ______________________________ Date: __________

Multiplying in Columns

Directions: Solve the Multiplications

1450 × 258	4750 × 475	1475 × 241	7896 × 120
3210 × 784	6520 × 784	7896 × 214	4785 × 124
2134 × 823	4756 × 124	1340 × 321	1235 × 365

Name: ____________________ Date: __________

Multiplying in Columns

Directions: Solve the Multiplications

3250 × 147	7854 × 241	9875 × 142	3201 × 201
3201 × 147	5478 × 235	4710 × 210	4750 × 240
1245 × 325	7845 × 321	4578 × 230	7854 × 342

Name: ______________________ Date: ____________

Multiplying in Columns

Directions: Solve the Multiplications

7845 × 533	6523 × 142	2310 × 478	2451 × 142
3214 × 450	4782 × 245	4120 × 213	2140 × 241
3214 × 412	4201 × 230	2310 × 203	7845 × 214

Name: ______________________ Date: __________

Multiplying in Columns

Directions: Solve the Multiplications

4578 × 541	1420 × 254	4710 × 245	2145 × 120
1450 × 120	4210 × 124	1452 × 123	4510 × 230
1202 × 230	1420 × 253	4780 × 142	4750 × 123

Name: ______________________ Date: __________

Multiplying in Columns

Directions: Solve the Multiplications

4170 × 124	1475 × 145	4782 × 124	4400 × 333
5500 × 555	8770 × 880	4512 × 210	3240 × 451
1474 × 215	1245 × 231	7788 × 550	4478 × 223

Name: ______________________ Date: ____________ 20

Multiplying in Columns

Directions: Solve the Multiplications

2223 × 141	7878 × 200	3265 × 412	1124 × 120
4545 × 112	4745 × 231	1122 × 111	3256 × 222
3356 × 333	7896 × 555	7778 × 777	2520 × 254

Name: ______________________ Date: ____________

Multiplying in Columns

Directions: Solve the Multiplications

5201 × 320	4752 × 258	3257 × 147	1457 × 325
1245 × 145	2134 × 120	2310 × 230	3780 × 731
1245 × 120	3940 × 941	7319 × 193	8430 × 194

Name: ____________________ Date: __________

Multiplying in Columns

Directions: Solve the Multiplications

3214 × 245	3214 × 213	1452 × 321	2154 × 204
1452 × 124	3124 × 846	6734 × 163	3461 × 124
3941 × 210	3110 × 230	4710 × 210	3210 × 210

Name: ______________________ Date: ____________

Multiplying in Columns

Directions: Solve the Multiplications

4210 × 120	3240 × 231	3692 × 142	2450 × 321
6980 × 159	9510 × 257	7563 × 145	7896 × 314
1234 × 457	14523 × 412	4532 × 214	2451 × 326

Name: ______________________________ Date: ____________

Multiplying in Columns

Directions: Solve the Multiplications

3698 × 214	3210 × 145	7854 × 213	3240 × 951
1452 × 214	3245 × 145	2450 × 124	3541 × 124
4521 × 124	3214 × 214	2456 × 789	5566 × 321

Name: ______________________ Date: ____________

Multiplying in Columns

Directions: Solve the Multiplications

1452 × 412	4785 × 142	4541 × 951	4563 × 258
1452 × 124	1452 × 142	4578 × 124	4750 × 451
1245 × 142	4521 × 120	4125 × 123	1452 × 123

Name: ______________________ Date: __________

Multiplying in Columns

Directions: Solve the Multiplications

4562 × 123	4785 × 321	4580 × 145	7895 × 214
7845 × 125	4520 × 780	4750 × 145	1452 × 124
4521 × 145	4521 × 321	4587 × 123	4587 × 125

Name: ____________________ Date: __________

Multiplying in Columns

Directions: Solve the Multiplications

4578 × 147	4578 × 152	4780 × 145	4520 × 147
5841 × 145	4578 × 345	6254 × 147	4785 × 241
1478 × 185	1452 × 245	7350 × 140	7845 × 840

Name: ______________________ Date: __________

Multiplying in Columns

Directions: Solve the Multiplications

4780×124	7845×235	4578×245	3652×285
8524×245	3570×654	7895×365	7865×569
8540×243	1954×951	7532×164	7624×243

Name: ____________________ Date: __________

Multiplying in Columns

Directions: Solve the Multiplications

4521 × 213	3214 × 245	1420 × 452	7541 × 452
1452 × 245	7854 × 214	7896 × 321	7854 × 124
3260 × 458	7865 × 245	3210 × 214	3210 × 231

Name: ______________________ Date: __________

Multiplying in Columns

Directions: Solve the Multiplications

4120 × 321	3652 × 421	5860 × 478	7895 × 214
4785 × 245	3520 × 142	1230 × 456	8745 × 214
7840 × 326	6350 × 324	3256 × 214	3698 × 325

Name: ______________________ Date: __________

Multiplying in Columns

Directions: Solve the Multiplications

1212 × 112	3232 × 232	3326 × 110	1124 × 221
2510 × 258	8541 × 147	2147 × 369	9632 × 254
3214 × 245	8790 × 987	2654 × 321	2454 × 214

Name: ______________________________ Date: ____________

Multiplying in Columns

Directions: Solve the Multiplications

7854 × 213	1235 × . 321	7896 × 321	7854 × 452
3254 × 365	7523 × 349	3589 × 869	3214 × 201
3210 × 254	2144 × 778	8879 × 325	1452 × 320

Name: ______________________ Date: __________

Multiplying in Columns

Directions: Solve the Multiplications

1452 × 120	4562 × 156	5641 × 240	4652 × 150
9875 × 563	9842 × 560	4652 × 654	8250 × 875
6451 × 123	4562 × 260	4650 × 456	4165 × 654

Name: ______________________________ Date: ____________

Multiplying in Columns

Directions: Solve the Multiplications

2650 × 654	6532 × 642	5462 × 540	8720 × 321
4652 × 250	5478 × 451	1452 × 564	4680 × 465
4980 × 840	5648 × 872	5168 × 780	8502 × 546

Name: ______________________ Date: __________

Multiplying in Columns

Directions: Solve the Multiplications

4820 × 321	6540 × 489	7892 × 654	8972 × 874
7892 × 546	7890 × 897	7820 × 564	7890 × 560
5892 × 852	8542 × 258	3654 × 451	7854 × 245

Name: ______________________________ Date: ____________

Multiplying in Columns

Directions: Solve the Multiplications

6542 × 456	4650 × 120	7854 × 321	4658 × 124
1245 × 325	3214 × 478	7850 × 364	3250 × 147
1482 × 421	4567 × 124	4512 × 325	3265 × 452

Name: ____________________ Date: ____________

Multiplying in Columns

Directions: Solve the Multiplications

5640 × 875	2340 × 124	3698 × 452	3240 × 785
3214 × 546	2356 × 152	1200 × 230	3212 × 200
3256 × 124	8765 × 450	7850 × 123	5432 × 214

Name: ______________________ Date: __________

Multiplying in Columns

Directions: Solve the Multiplications

5487 × 125	4780 × 325	6980 × 458	3250 × 145
1325 × 125	3212 × 457	7890 × 587	3125 × 214
3256 × 124	3250 × 321	3254 × 210	3241 × 250

Name: ____________________ Date: ____________

Multiplying in Columns

Directions: Solve the Multiplications

1547 × 745	7812 × 145	2145 × 254	2132 × 158
4578 × 321	1245 × 215	2130 × 256	3250 × 325
3256 × 231	7810 × 235	4512 × 230	2123 × 231

Name: ______________________ Date: ____________

Multiplying in Columns

Directions: Solve the Multiplications

1254 × 124	2350 × 325	2354 × 289	9865 × 689
8945 × 123	5680 × 780	3265 × 123	3250 × 780
1245 × 235	4598 × 231	2350 × 487	7820 × 123

Name: ______________________ Date: ____________

Multiplying in Columns

Directions: Solve the Multiplications

2359 × 512	8426 × 215	8457 × 842	6535 × 620
2145 × 236	4587 × 458	9637 × 142	4587 × 123
1258 × 590	2520 × 326	3550 × 120	3265 × 256

Name: ______________________ Date: ____________

Multiplying in Columns

Directions: Solve the Multiplications

6859 × 953	5789 × 486	5478 × 153	2145 × 241
8632 × 298	8650 × 312	3250 × 265	4580 × 245
5462 × 235	2560 × 245	3256 × 325	5462 × 214

Name: ______________________ Date: ____________

Multiplying in Columns

Directions: Solve the Multiplications

6235 × 214	5478 × 518	6859 × 574	6587 × 620
2543 × 145	2365 × 214	2365 × 562	2103 × 256
6598 × 256	3256 × 326	5870 × 659	3256 × 128

Name: ______________________ Date: __________

Multiplying in Columns

Directions: Solve the Multiplications

3545 × 426	6803 × 425	5692 × 324	5689 × 325
2356 × 125	6598 × 214	3265 × 125	5468 × 569
3256 × 258	8523 × 365	3266 × 695	3214 × 247

Name: ______________________ Date: ____________

Multiplying in Columns

Directions: Solve the Multiplications

4752 × 150	6532 × 145	1254 × 128	9850 × 236
3698 × 247	3369 × 478	9987 × 235	4780 × 245
12450 × 365	4870 × 256	4521 × 326	1245 × 320

Name: ____________________ Date: __________

Multiplying in Columns

Directions: Solve the Multiplications

1221 × 125	12655 × 230	4587 × 326	8965 × 680
4578 × 845	8950 × 325	8560 × 478	7850 × 879
8745 × 214	7540 × 325	7840 × 124	3210 × 145

Name: ______________________ Date: ____________

Multiplying in Columns

Directions: Solve the Multiplications

9860 × 351	1325 × 512	3652 × 352	3562 × 353
9875 × 125	4578 × 865	9568 × 685	5924 × 563
4521 × 123	7896 × 982	9856 × 896	3255 × 665

Name: ____________________ Date: __________

Multiplying in Columns

Directions: Solve the Multiplications

5628 × 147	9845 × 458	3250 × 258	3690 × 478
9850 × 420	8751 × 587	5698 × 250	3562 × 256
4521 × 365	7540 × 145	1245 × 145	3254 × 210

Name: ______________________ Date: ____________

Multiplying in Columns

Directions: Solve the Multiplications

4578 × 214	3256 × 124	4785 × 342	4521 × 124
1452 × 120	4780 × 125	1452 × 124	4520 × 120
3258 × 241	3562 × 458	3256 × 124	1234 × 340

Name: ______________________ Date: ____________

Multiplying in Columns

Directions: Solve the Multiplications

4521 × 360	4785 × 123	4587 × 125	4521 × 123
4521 × 125	1230 × 124	4521 × 120	3214 × 210
9850 × 952	3692 × 253	1236 × 352	3652 × 256

Made in the USA
Monee, IL
05 November 2026

67379323R10028